MACHINES

John Williams

Illustrated by
Malcolm S. Walker

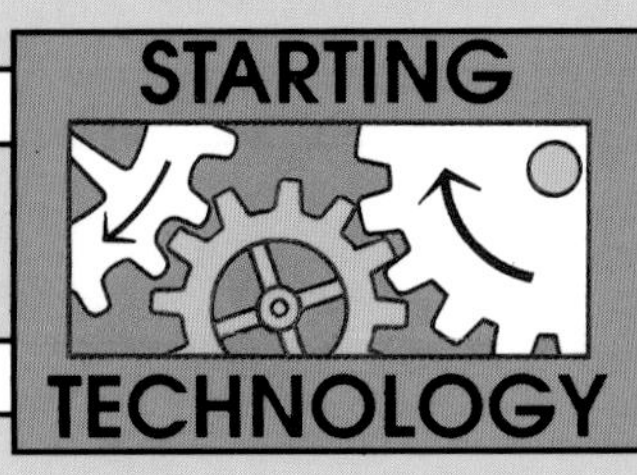

Titles in this series

AIR
COLOUR AND LIGHT
ELECTRICITY
FLIGHT
MACHINES
TIME
WATER
WHEELS

Words printed in **bold** appear in the glossary on page 30

First published in 1991 by
Wayland (Publishers) Ltd
61 Western Road, Hove
East Sussex BN3 1JD, England

Editor: Anna Girling
Designer: Kudos Design Services

British Library Cataloguing in Publication Data
Williams, John
Machines.
1. Machinery
I. Title II. Series
621.8
ISBN 0 7502 0025 1

Typeset by Kudos Editorial and Design Services, Sussex, England
Printed in Italy by Rotolito Lombarda S.p.A.
Bound in Belgium by Casterman S.A.

CONTENTS

STARTING TECHNOLOGY

SIMPLE MACHINES

Machines help us to do work. They can be very simple, like tin-openers. They may be very large and complicated, like the machines used in **factories**.

All machines are based on simple ideas. **Levers**, **screws** and **pulley wheels** are all simple machines. However big they are, machines are just a lot of simple ideas, like these, linked together.

In this factory the big, complicated machines are controlled by computers. They do not need lots of people to look after them.

Making a moving cat

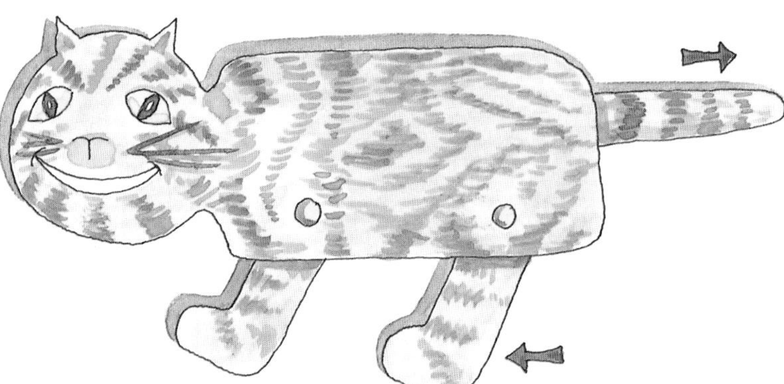

You will need:

Card
Pencils
Paper fasteners
Scissors
Paints

1. Draw the shape of a cat's head and body on card and cut it out.

2. Draw two legs and a tail, making sure they are at least 2 cm wide. The legs should be about 10 cm long and the tail about 25 cm long.

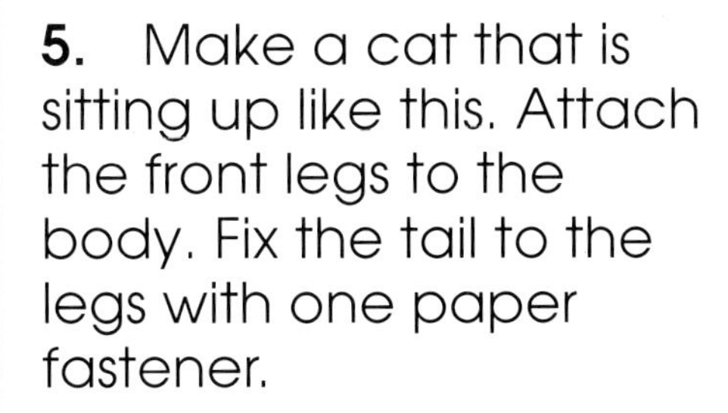

3. Fix the *legs only* to the back of the cat's body with paper fasteners. Fix the tail to the top of the legs, but not to the body, with fasteners. Now you can move the tail and the legs will also move.

4. Paint the cat's face and body on the other side of the card.

5. Make a cat that is sitting up like this. Attach the front legs to the body. Fix the tail to the legs with one paper fastener.

MAKING SIGNALS

Today, railways use very complicated signal lights to make sure all the trains are in the right place at the right time. In the past, railway signals were much simpler. They used a system of levers to move a signal up and down. Do you have a model train set? Does it have old-style signals?

These are old-style railway signals. Most railways now use coloured lights for signals.

Making an old-style railway signal

You will need:

White card
A pencil and ruler
Four paper fasteners
Scissors
Red paint

1. Cut four strips of card. The main stand should be about 20 cm long by 3 cm wide, the bar 18 cm by 2 cm, the signal 11 cm by 3 cm, and the handle 9 cm by 3 cm.

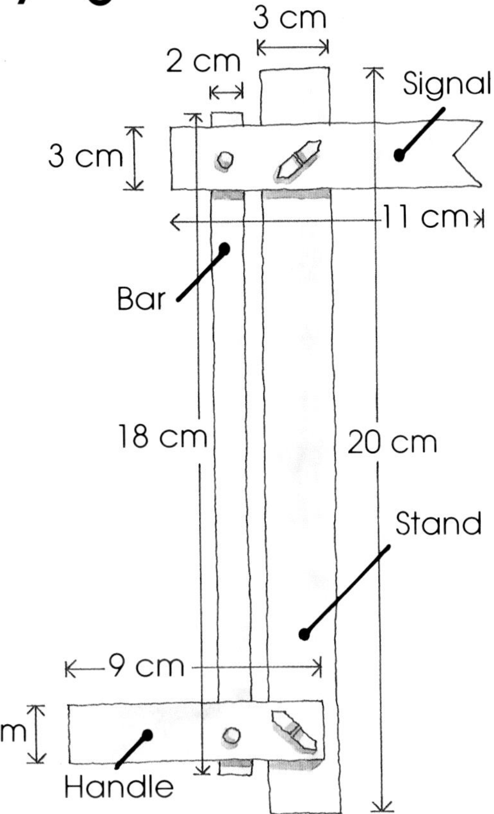

2. First fix the signal and handle to each end of the bar with paper fasteners. Then fix the signal and handle to the stand with fasteners as shown in the diagram above. Now you can move the handle to make the signal go up and down.

3. Cut a V-shape in the end of the signal and paint a red stripe on it to make it look like a real signal.

ARMS AND LEGS

Our bodies work like machines. Bones in our arms and legs work like levers. The **joints** act as **fulcrums**. When you nod your head or open and shut your mouth you are working a lever.

This funny robot was made to be a character in a film. It has mechanical arms which work rather like human ones. Can you see the joints and levers?

Making a model arm

You will need:

Card
A pencil and ruler
Paper fasteners
Sticky tape
A rubber band
Scissors

1. Cut out two strips of card, 10 cm long by 2 cm wide. Cut out the shape of a hand and stick it to the end of one of the strips.

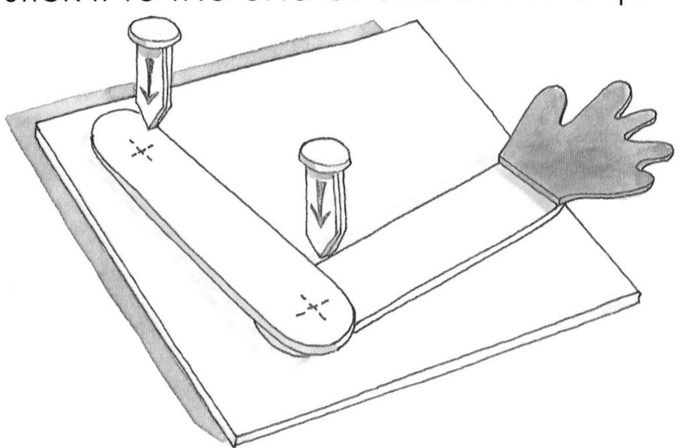

2. Join the strips together at the 'elbow' with a paper fastener and use the same fastener to fix the arm to a large piece of card. Attach the top of the arm to the card with another fastener.

3. Fix the rubber band to the centre of each piece of card with sticky tape. Try moving the lower part of the arm up and down. Hang something quite heavy on the hand. What happens?

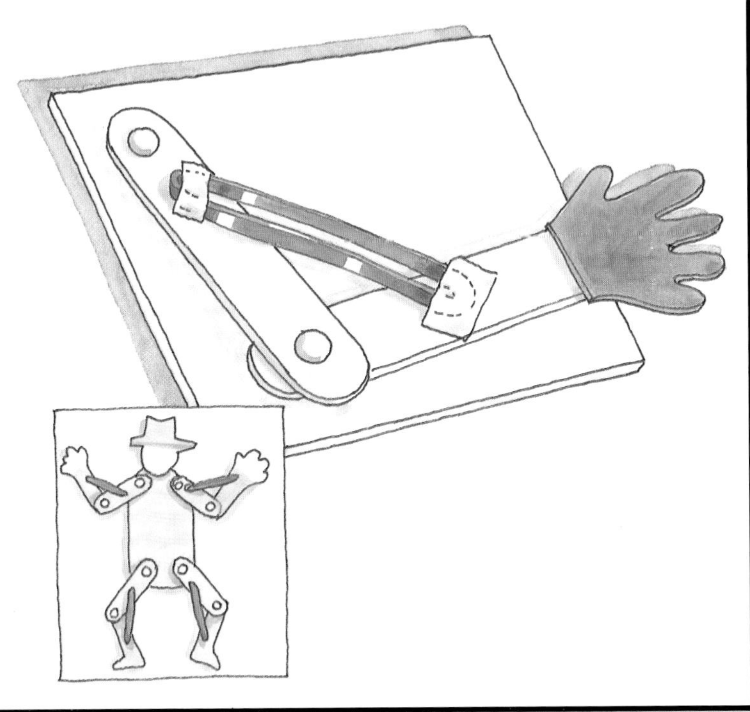

4. You can make a model 'skeleton' like this, with two arms and two legs.

STRETCHING ARMS

Making a mechanical stretching arm

You will need:

Card
A pencil and ruler
Paper fasteners
Scissors
Glue

1. Cut out eight strips of card 15 cm long and 2 cm wide.

2. Join the strips together with paper fasteners. Fasten them exactly in the middle and at each end.

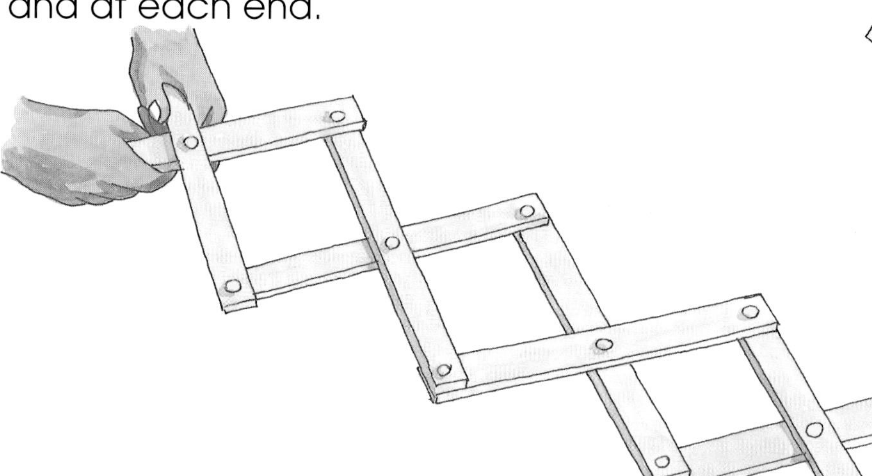

3. At one end of the arm, glue two triangular-shaped 'fingers'. By moving the pieces of card at the other end, you can make the fingers move. You can also make the arm stretch out.

Further work

You can use wooden spatulas or lollipop sticks, instead of card, to make the arm. They make very good stretching arms. Ask an adult to help you make the holes for the paper fasteners.

This lifting platform is made up of lots of simple levers, one above another.

CATAPULTS

Catapults are machines used to make objects fly through the air. Long ago they were used in battles, to hurl stones.

Making a catapult

You will need:

A block of wood
Two nails
A small hammer
A rubber band
A plastic spoon
A clothes peg
Card
Drawing pins
Glue
Scissors
Plasticine

WARNING: Ask an adult to help you with the hammer and nails.

1. Hammer the nails firmly into the wooden block, about 10 cm apart. There should be at least 2 cm of each nail showing above the wood.

2. Stretch a twisted rubber band between the nails. Push the handle of the spoon through the twisted rubber band.

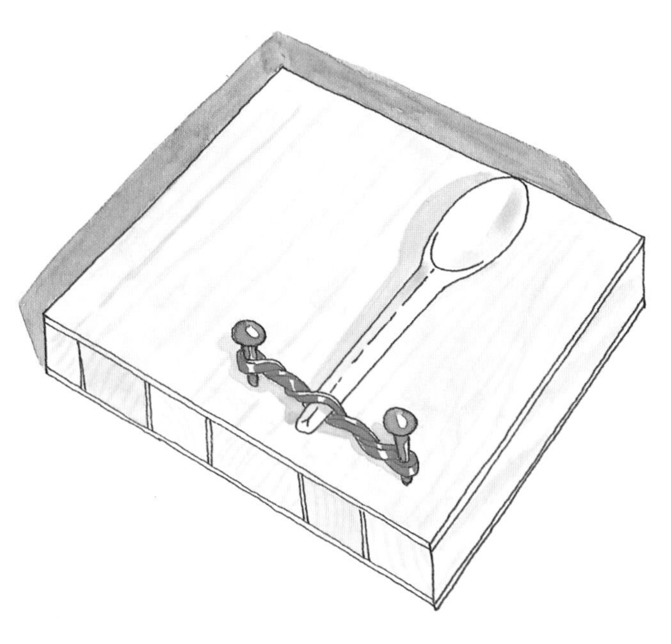

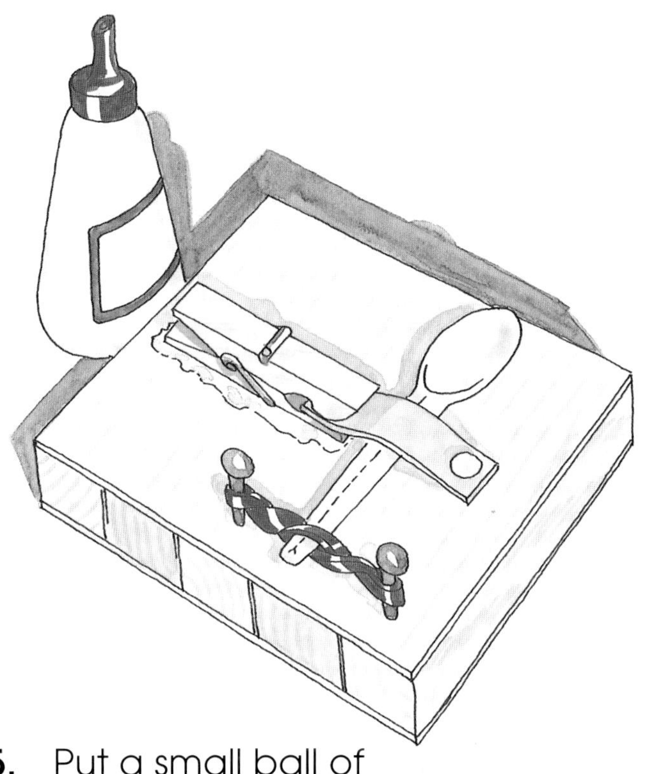

3. You will need to hold the spoon in place with a strip of card. To do this, first glue the clothes peg to the wooden block, at **right angles** to the spoon.

4. Cut out a strip of card. Use the clothes peg to grip one end of the strip and pin the other end to the block, on the other side of the spoon, so that the card goes across the handle.

5. Put a small ball of plasticine into the bowl of the spoon. Open the clothes peg to release the card. How far does the catapult shoot the plasticine?

WARNING: Always be careful when firing your catapult. Never fire heavy or sharp objects. Make sure you will not hit anybody or damage anything.

6. Try your catapult with more or less twists in the rubber band. Does it make any difference? When testing your catapult, always use the same amount of plasticine – otherwise the tests will not be fair.

DIGGERS

Diggers and **bulldozers** are used to move soil and rocks to prepare the ground for new roads and buildings. Some diggers have enormous buckets on them. Have you ever seen a digger on a building site?

A bulldozer can push large amounts of rocks and stones from place to place.

Making a mechanical digger

You will need:

A cardboard box about 20 cm long
Card
Paper fasteners
Four cardboard circles for wheels
Glue
Scissors

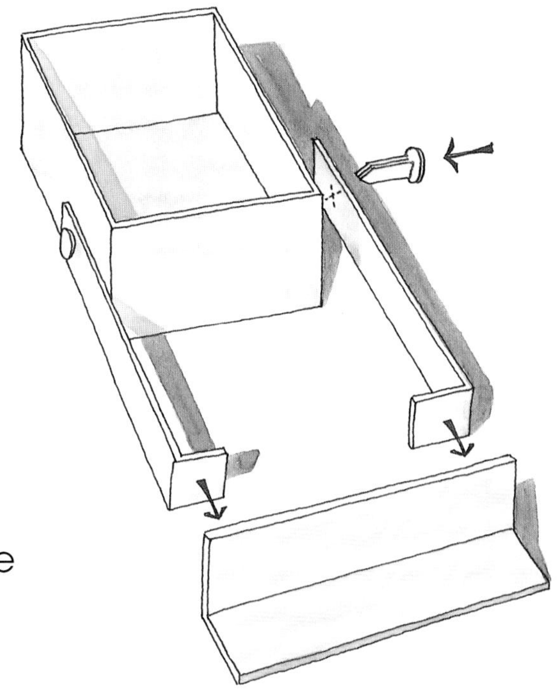

1. Cut out a rectangle of card for the scoop. It should be the same width as your box. Fold it to make a base and a back.

2. Cut out two strips of card, about 20 cm long and 2 cm wide, to make the main arms for the digger. Make a fold in each strip, about 5 cm from the end, and glue the folded ends to the scoop. Fix the other ends of the arms to the box with paper fasteners.

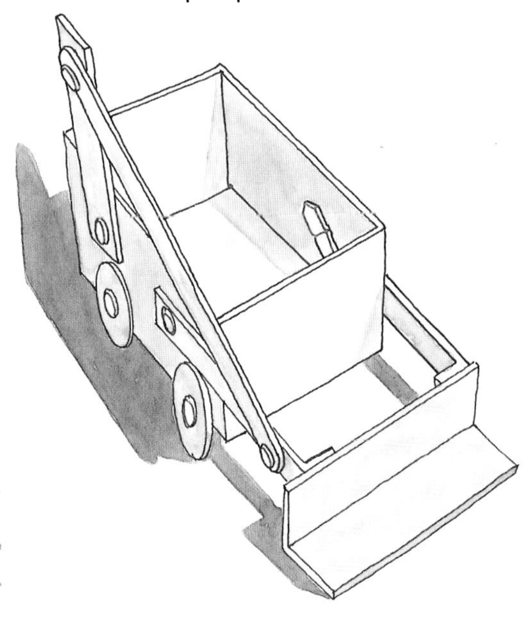

3. Cut a piece of card about 12 cm long for the handle. With a paper fastener, fix this to the side of the box, at the back of your digger.

4. Cut out a strip of card for the lever bar, long enough to reach from the top of the handle to the digger arm. Attach it at both ends with fasteners. Pull the handle and the scoop will move up and down.

5. Fix wheels to the box with paper fasteners.

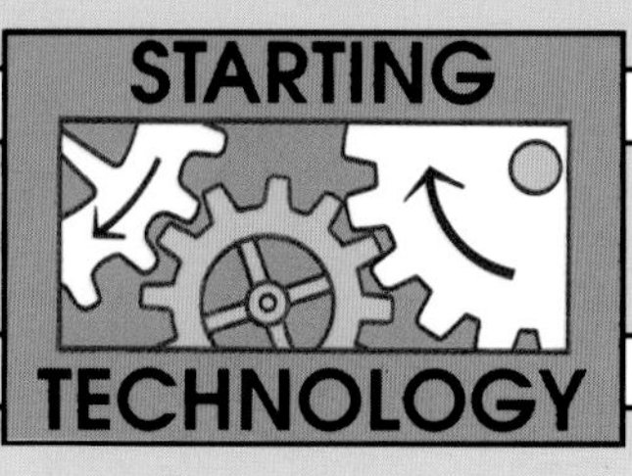

PULLEY WHEELS

Pulley wheels are used for lifting heavy objects. A pulley is a wheel with a rope round it. By pulling down on one end of the rope you can lift a heavy load on the other end. Do you think pulling down a heavy object is easier than lifting it up?

Using a pulley wheel

You will need:

A large matchbox
Paperclips
Sticky tape
Cotton thread
Nylon fishing line
Rubber bands
Two hooks
A cotton reel
A garden cane
A plastic beaker
A stand and clamp
Plasticine

1. You will need to stretch a piece of fishing line between two table legs, so that it slopes gently. Cut the length of line you need and tie a rubber band to each end. Attach the bands to the table legs with hooks or sticky tape. Make sure the line is tight.

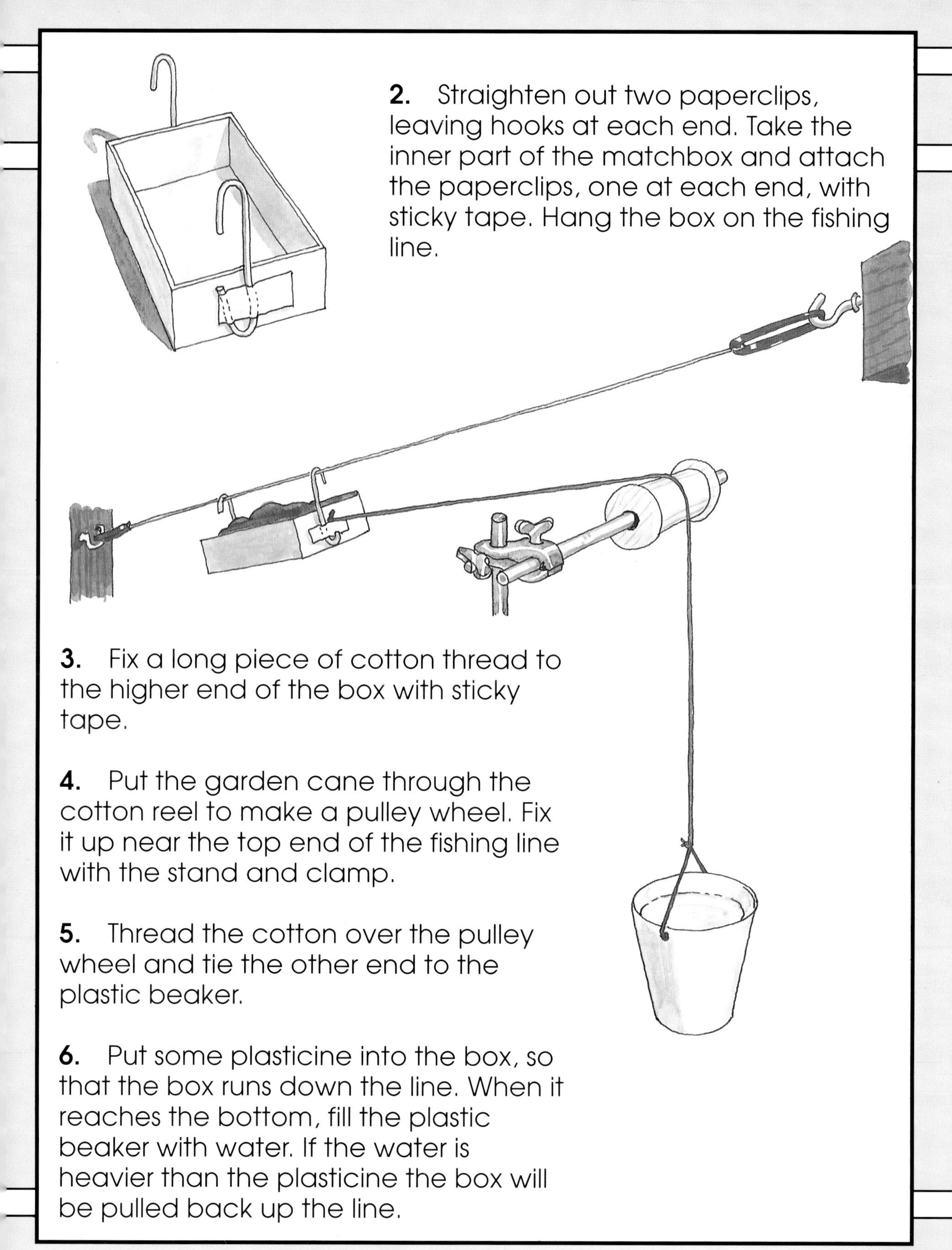

2. Straighten out two paperclips, leaving hooks at each end. Take the inner part of the matchbox and attach the paperclips, one at each end, with sticky tape. Hang the box on the fishing line.

3. Fix a long piece of cotton thread to the higher end of the box with sticky tape.

4. Put the garden cane through the cotton reel to make a pulley wheel. Fix it up near the top end of the fishing line with the stand and clamp.

5. Thread the cotton over the pulley wheel and tie the other end to the plastic beaker.

6. Put some plasticine into the box, so that the box runs down the line. When it reaches the bottom, fill the plastic beaker with water. If the water is heavier than the plasticine the box will be pulled back up the line.

CRANES

Cranes use pulleys to lift big, heavy objects. Enormous cranes are used to build tall modern office blocks or load huge containers on to ships.

Cranes are used to help build very tall buildings. Look how tall this crane is.

Making a crane

You will need:

Card
Scissors
Wooden sticks
Cotton thread
A small cardboard box
Glue
A hook or magnet

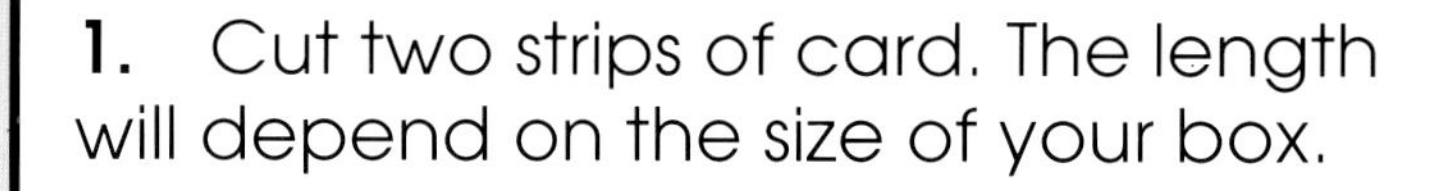

1. Cut two strips of card. The length will depend on the size of your box.

2. Push a short piece of stick through the end of both strips. Push a long piece through the other ends. Fix the sticks firmly with glue.

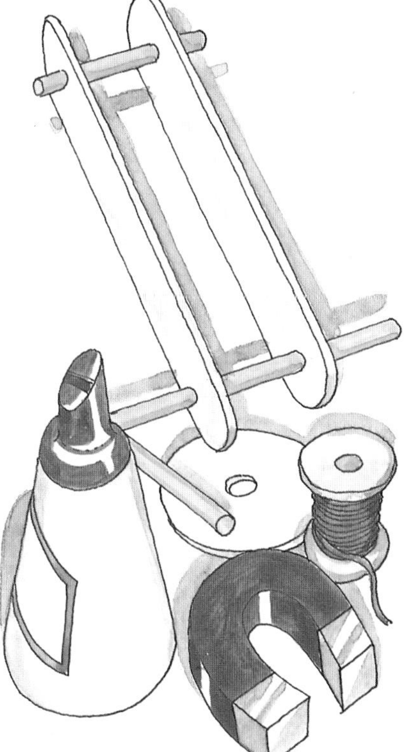

3. Push the ends of the longer stick through the sides of the box. The strips of card should rest on the front edge.

4. Tie a piece of cotton thread to another stick and push the ends of the stick through the sides of the box, at the back of the crane. Thread the cotton over the short stick at the front of the crane.

5. Make a handle for winding up the cotton, from card, glue and a short piece of stick. Tie the hook or magnet to the other end of the thread and try to pick something up. Put some plasticine at the back of the crane to weigh it down.

BALLOON POWER

Making a tip-up lorry

You will need:

A balloon
Plastic tubing
Sticky tape
A plastic squeezy bottle
Scissors
A shoebox with a lid
Smaller boxes
Card
Paper fasteners

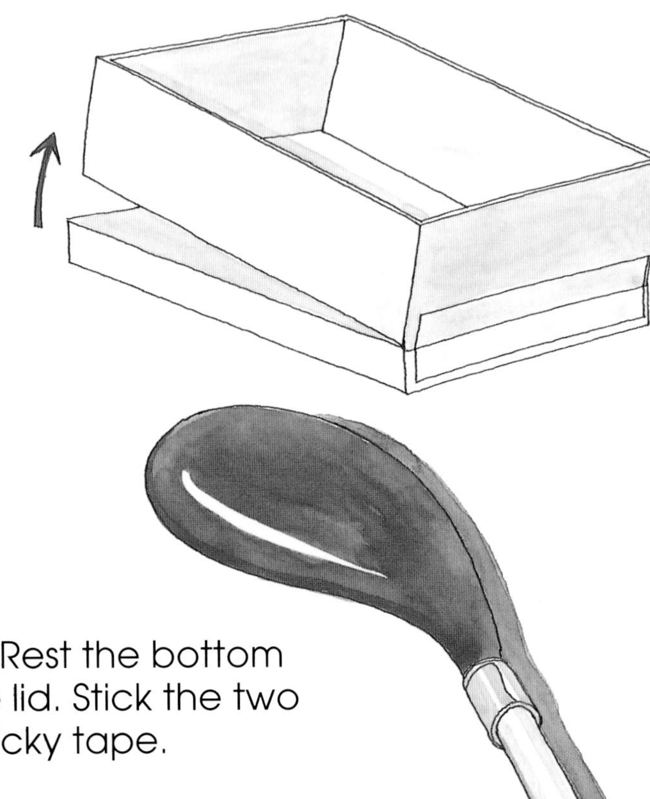

1. Take the lid off the box. Rest the bottom of the box on the top of the lid. Stick the two together at one end with sticky tape.

2. Cut the lip off the balloon. Stretch the mouth of the balloon tightly round the end of a piece of tubing. Fix it firmly with sticky tape.

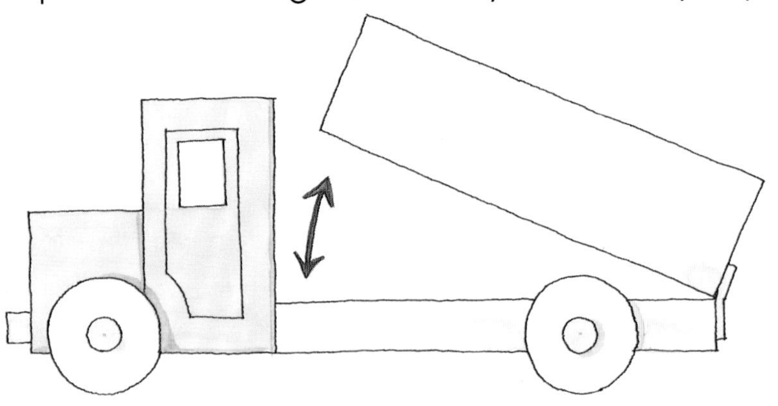

3. Attach other boxes to the other end of the lid for the driver's cab and engine. Use circles of card for wheels and fix them to the lorry with paper fasteners.

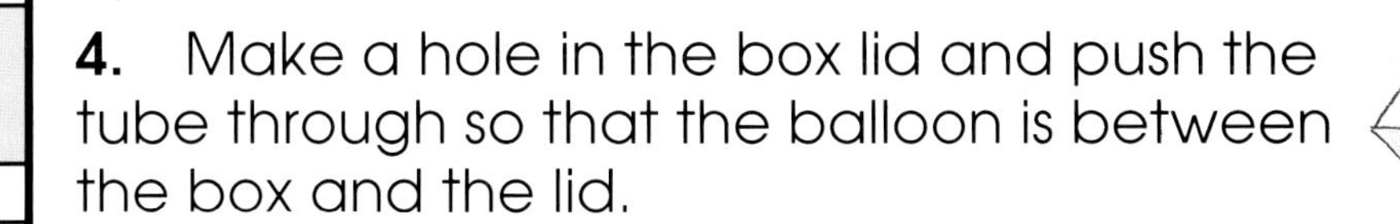

4. Make a hole in the box lid and push the tube through so that the balloon is between the box and the lid.

5. Fit the other end of the tube tightly on to the nozzle of the plastic bottle. Squeeze the bottle to blow up the balloon. It helps to make a hole in the end of the bottle. To keep air in the balloon, put your finger over the hole.

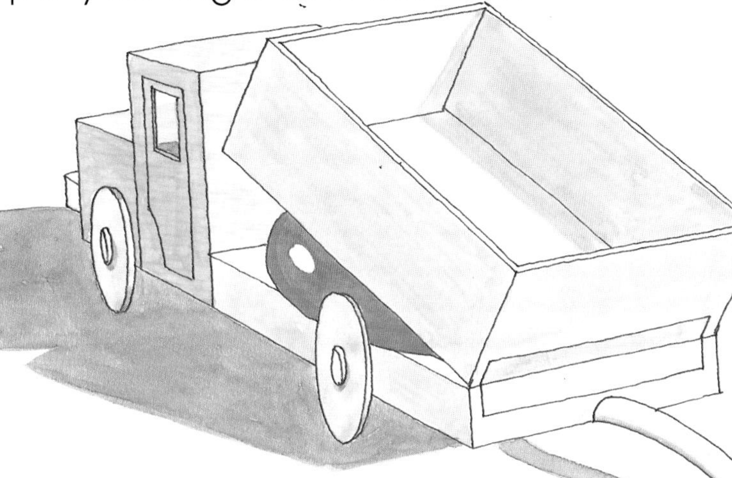

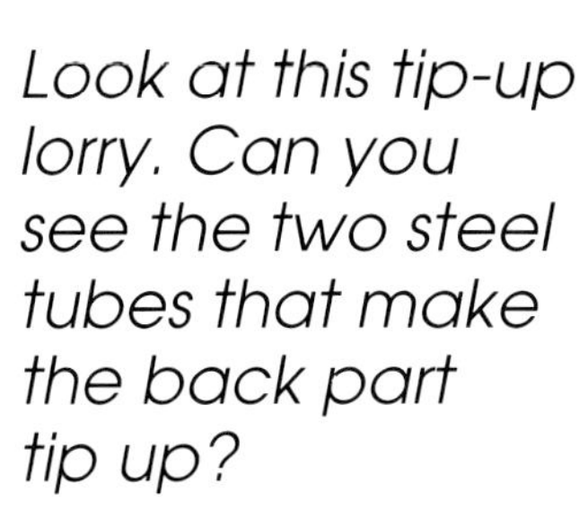

Look at this tip-up lorry. Can you see the two steel tubes that make the back part tip up?

DRAGONS

Making a moving dragon's mouth

You will need:

A balloon
Plastic tubing
Sticky tape
A plastic squeezy bottle
Scissors
A shoebox with a lid
Card
Glue
Egg boxes
String
Paints

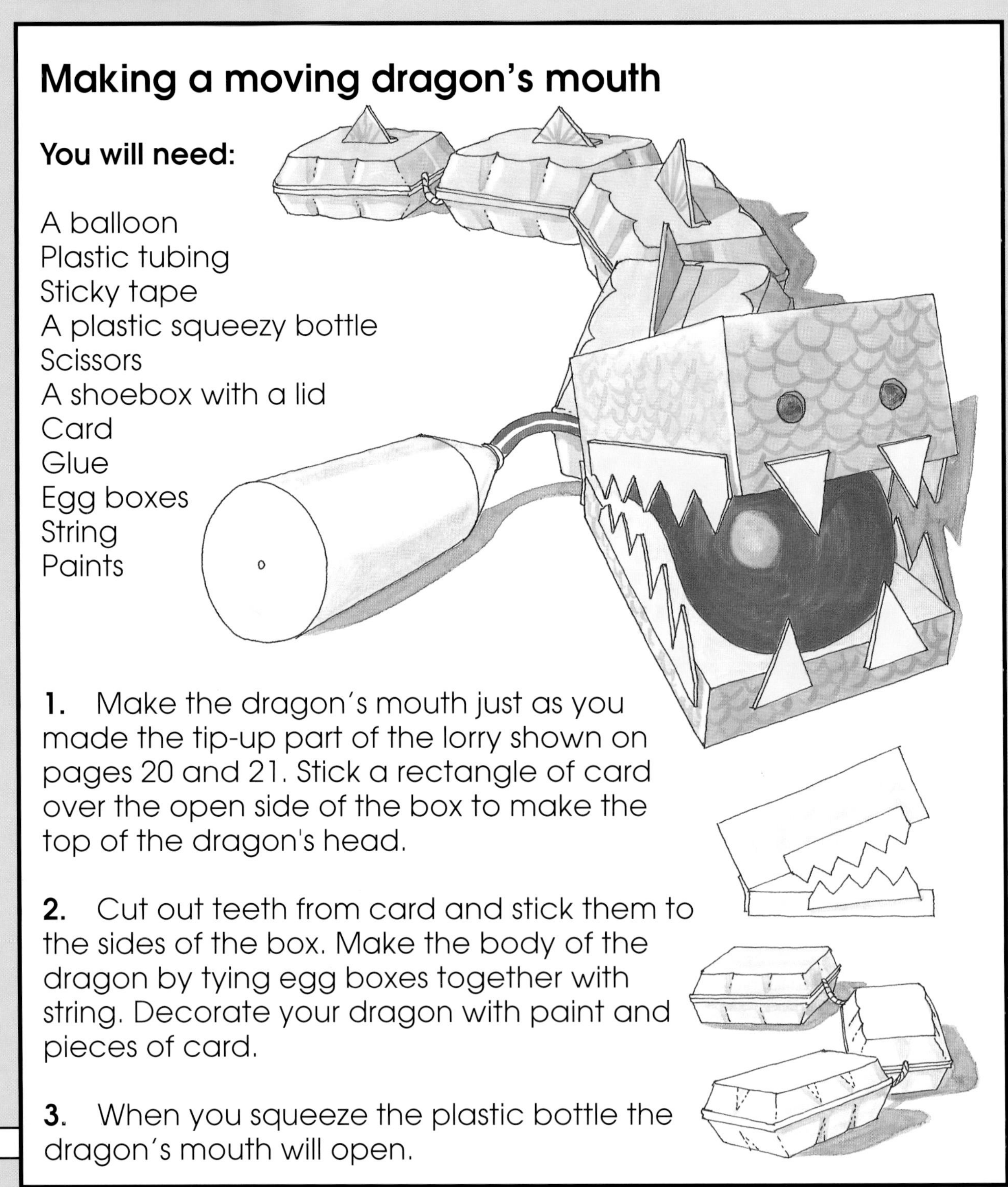

1. Make the dragon's mouth just as you made the tip-up part of the lorry shown on pages 20 and 21. Stick a rectangle of card over the open side of the box to make the top of the dragon's head.

2. Cut out teeth from card and stick them to the sides of the box. Make the body of the dragon by tying egg boxes together with string. Decorate your dragon with paint and pieces of card.

3. When you squeeze the plastic bottle the dragon's mouth will open.

Making the dragon's eyes

WARNING: Never use mains electricity from your house. It is very dangerous.

You will need:

Two electric bulbs with bulb holders
Insulated copper wires
A 4.5-volt **battery**

1. Use the wires to attach the bulb holders and battery together to make an **electric circuit**. The bulbs will light up.

2. Fix the bulb holders to the top of the dragon's head with sticky tape to make eyes.

In China and other parts of the world people hold dragon festivals. They make dragons like this one and dance with them in the street.

SWITCHES

Televisions, radios and electric lights all have switches for turning them on and off. Look around you at home and at school for other machines with switches on them. But remember: never play around with switches because electricity can be very dangerous.

There are many switches in an aeroplane cockpit. See how many you can count in this picture. The pilot has to know exactly what each one is for.

Making a see-saw switch

You will need:

Card
A cardboard box
Paper fasteners
Kitchen foil
A 4.5-volt battery
Insulated copper wire
A bulb and bulb holder
Scissors

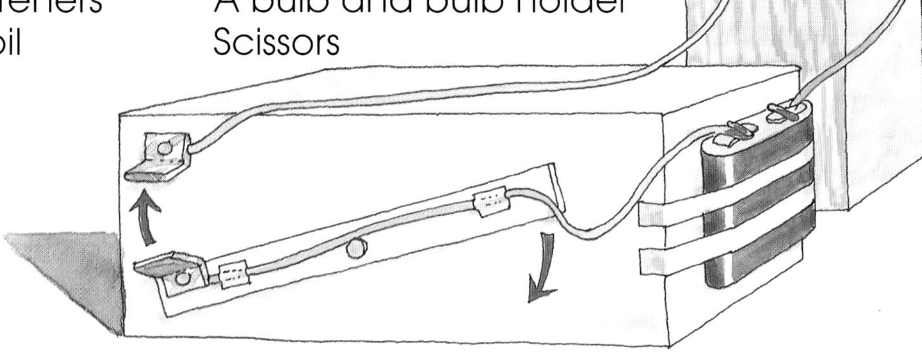

1. Cut a strip of card about 12 cm long. Fix it to the side of the box with a paper fastener to make a kind of see-saw. Make sure the fastener is in the middle of the piece of card.

2. Fix a flap of kitchen foil to one end of the card with a paper fastener. Fasten another flap to the box, just above.

3. Wind the end of a piece of wire around each of these fasteners, between the foil and the card.

4. Strap the battery to the end of the box with sticky tape. Join the wire on the see-saw to the battery and the wire on the top flap to one of the screws on the bulb holder. Join the battery to the other screw on the bulb holder with another piece of wire.

5. Push the see-saw so that the metal flaps touch. Does the bulb light up?

WARNING: Never use mains electricity from your house. It is very dangerous.

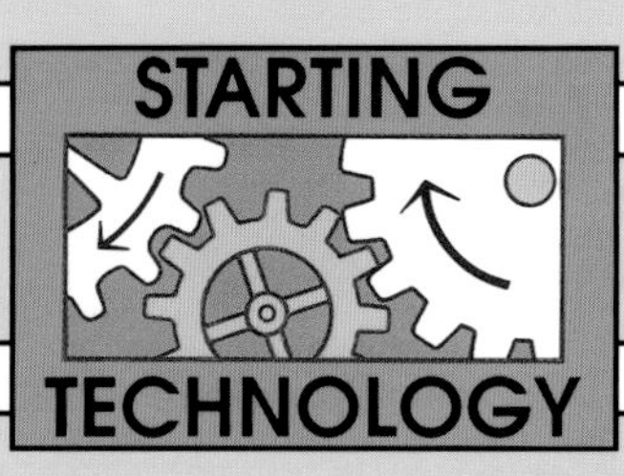

A SWITCH MACHINE

Making an automatic switch machine

You will need:

A large cardboard box
A marble
Card
Cotton thread
A plastic cup
Paper fasteners
A wooden stick
Drinking straws
Sticky tape
Kitchen foil
Insulated copper wire
A bulb and bulb holder
A 4.5-volt battery
Scissors
Glue

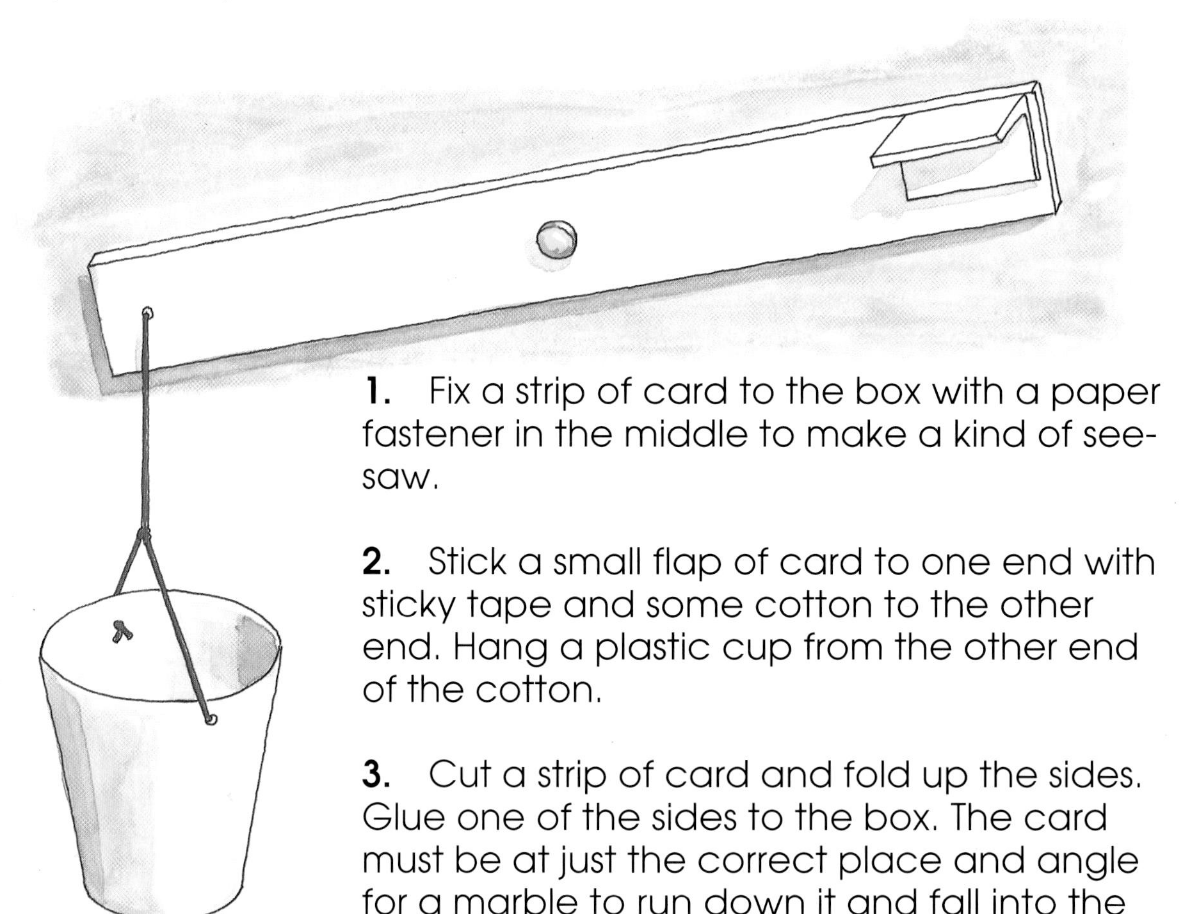

1. Fix a strip of card to the box with a paper fastener in the middle to make a kind of see-saw.

2. Stick a small flap of card to one end with sticky tape and some cotton to the other end. Hang a plastic cup from the other end of the cotton.

3. Cut a strip of card and fold up the sides. Glue one of the sides to the box. The card must be at just the correct place and angle for a marble to run down it and fall into the cup.

4. Stick a straw to the box with sticky tape, directly above the flap of card.

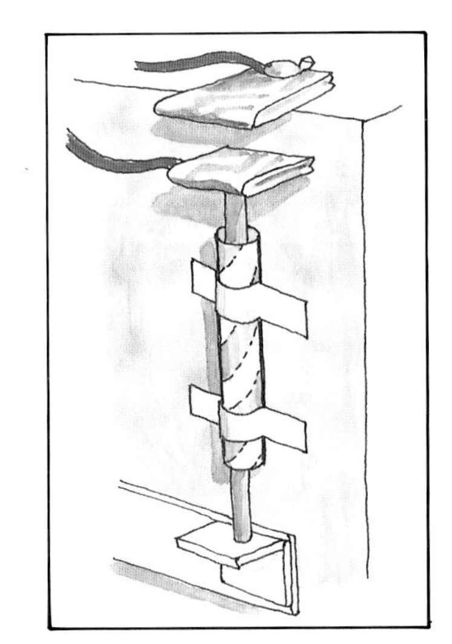

5. Put the wooden stick through the straw so that it rests on the flap of card. Glue a folded piece of kitchen foil to the other end of the stick. Stick the end of a piece of wire between the folds of foil.

6. Attach a flap of foil to the box with a paper fastener, directly above the top of the stick. Use more wire to connect up the two foil flaps, the bulb and battery – just as shown on page 25.

7. Run a marble down the slope. How many different machines have to work before the bulb lights up?

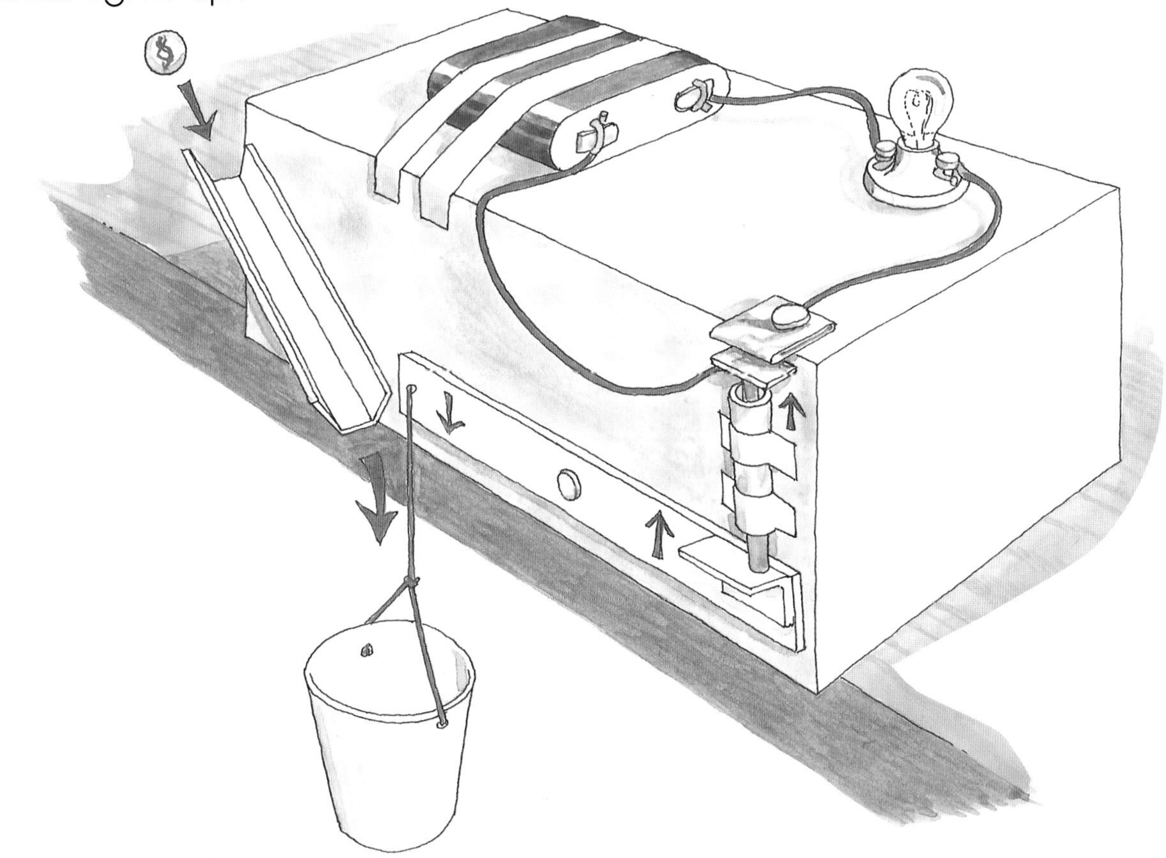

Notes for Parents and Teachers

SCIENCE

Machines, even simple ones, help us to do work. All levers are machines, and teachers and parents should be familiar with the three types of lever in order to help children understand the science involved in the models they are making. The table on the right shows the three types of lever.

DESIGN AND TECHNOLOGY

All the models shown in this book have been made by very young children – even in crowded classrooms. It is possible to make the projects class-based activities, but it is probably better for the models to be made by one group of children within the class. Other groups could be working at quieter activities. All the children should eventually have the opportunity to carry out the practical model-making work.

Safety should be stressed at all times. All the materials and tools suggested in this book are very simple and, when properly used, are quite safe. When experimenting with bulbs and batteries children should be made aware of the dangers of mains electricity.

The development of an understanding of the design process is very important. Once children have made the models in this book they should be encouraged to produce their own ideas for future designs. They should make drawings and plans from which they can build, develop and improve their own models.

Teachers should be aware of the possibilities of adapting the models in this book for use in control technology projects. Control technology is an important part of the National Curriculum for technology. Teachers should be able to find ways for children to use computers to control at least some of the working models shown in this book.

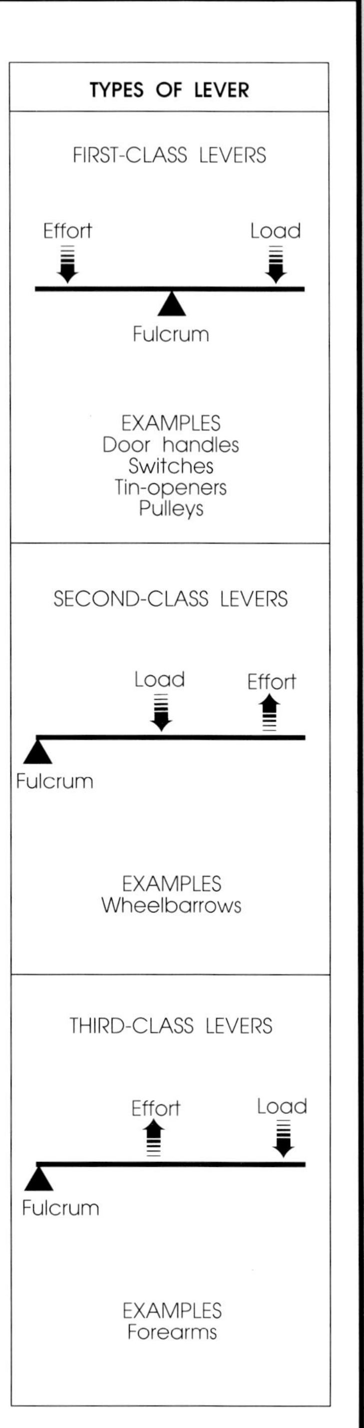

National Curriculum Attainment Targets

This book is relevant to the following Attainment Targets in the National Curriculum for science:

Attainment Target 1 (Exploration of science) The construction, development and testing of all the models in this book is relevant. Special attention should be paid to developing fair tests (Level 3).

Attainment Target 10 (Forces) The making and testing of mechanical objects is relevant to this Attainment Target. Forces – pushes or pulls – make things move or change the direction of objects that are already moving. Machines such as levers allow these forces to be properly used and multiply their effects. The force of friction is also involved in the moving parts.

Attainment Target 13 (Energy) All the working models in this book need a source of energy to make them move. Attention should also be paid to the way machines transfer energy.

The following Attainment Targets are included to a lesser extent:

Attainment Target 2 (The variety of life) Models of cats and human arms and legs can be used to introduce an element of technology into topics about living things.

Attainment Target 11 (Electricity and magnetism) The electric circuits in the chapters on switches and dragons are relevant.

Teachers should also be aware of the Attainment Targets covered in other National Curriculum documents – that is, those for design and technology, mathematics and language.

GLOSSARY

Battery A container with special chemicals in it that produce electricity.

Bulldozers Powerful tractors with blades on the front used for moving soil and rocks.

Electric circuit A loop of wires and objects connected up so that electricity will flow round it.

Factories Buildings containing complicated machines, where goods are made.

Fulcrum The point on which a lever balances or turns.

Joints The parts of the body where two or more bones are joined together.

Lever A bar used for lifting heavy weights. The bar balances or turns on a point, called a fulcrum. You press or pull on one end of the bar to lift the weight on the other end.

Pulley wheel A wheel with a rope round it, used for lifting things.

Right angle An angle is a corner where two lines meet. A right angle is a square corner, like the corner of a book or box.

Screw A kind of nail with a spiral groove around it so that it can be put in a hole and twisted, to fasten things together.

BOOKS TO READ

Cranes by R.J. Stephen (Franklin Watts, 1986)
Earthmovers by R.J. Stephen (Franklin Watts, 1986)
How Machines Work by Christopher Rawson (Usborne, 1988)
How Things Work by Martyn Bramwell (Usborne, 1989)
How Things Work by Robin Kerrod (Cherrytree Books, 1988)
Let's Look At Monster Machines by Andrew Langley (Wayland, 1990)

Picture acknowledgements
The publishers would like to thank the following for allowing their photographs to be reproduced in this book: Cephas Picture Library 24 (Nigel Blythe); Eye Ubiquitous 6 (J. Winkles); Hutchison Library 23 (Leslie Woodhead); Sefton Photo Library 11, 18; Topham Picture Library 8, 21; Zefa 4 (T. Horowitz), 14 (J. Pfaff). Cover photography by Zul Mukhida.

STARTING TECHNOLOGY

INDEX